# The Stars

*Written by*
Cynthia Pratt Nicolson

*Illustrated by*
Bill Slavin

Kids Can Press

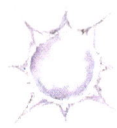

*For my mother, Bernice Pratt*

# Acknowledgments

Learning about the stars is a mind-stretching, sometimes mind-boggling business. I would like to thank Stan Shadick of the University of Saskatchewan for reading over my first draft and clearing up many points of confusion. His patient replies to my questions were invaluable and his Canadian Skywatcher's Trivia Calendar has inspired me with fascinating details about the wonders of the night sky.

My thanks also go to Grant Hill at the Dominion Astrophysical Observatory and to Colin Scarfe at the University of Victoria for their help in making this book as up to date and accurate as possible. Of course, any errors that may have crept into the text are my responsibility.

Ed Barker at the Manitoba Museum of Man and Nature planetarium and Sue Williams of the Bowen Island Astronomy Club have helped me go beyond the Big Dipper in my night sky navigation. Thanks to them both!

My appreciation for the work of everyone at Kids Can Press grows with every book we produce together. Thanks to Bill Slavin for the fun his illustrations bring to this book and to Val Wyatt for another expert job of editing. Finally, I would like to thank each member of my family for playing a starring role in my life.

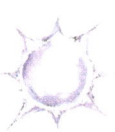

---

Text copyright © 1998 by Cynthia Pratt Nicolson
Illustrations copyright © 1998 by Bill Slavin

All rights reserved. No part of this publication may be reproduced, stored in a retrieval system or transmitted, in any form or by any means, without the prior written permission of Kids Can Press Ltd. or, in case of photocopying or other reprographic copying, a license from CANCOPY (Canadian Copyright Licensing Agency), 1 Yonge Street, Suite 1900, Toronto, ON, M5E 1E5.

Neither the Publisher nor the Author shall be liable for any damage which may be caused or sustained as a result of conducting any of the activities in this book without specifically following instructions, conducting the activities without proper supervision, or ignoring the cautions contained in the book.

Published in Canada by
Kids Can Press Ltd.
29 Birch Avenue
Toronto, ON  M4V 1E2

Published in the U.S. by
Kids Can Press Ltd.
85 River Rock Drive, Suite 202
Buffalo, NY  14207

NELVANA

Kids Can Press is a Nelvana company

Edited by Valerie Wyatt
Text design by Marie Bartholomew
Page layout and cover design by Esperança Melo

**Photo Credits**
All photos used courtesy of NASA.

Printed in Hong Kong by Wing King Tong Co. Ltd.

CM 98   0 9 8 7 6 5 4 3 2 1
CM PA 99   0 9 8 7 6 5 4 3 2 1

**Canadian Cataloguing in Publication Data**

Nicolson, Cynthia Pratt
    The stars

(Starting with space)
Includes index.
ISBN 1-55074-524-7 (bound)    ISBN 1-55074-659-6 (pbk.)

1. Stars — Juvenile literature.  I. Slavin, Bill.
II. Melo, Esperança.  III. Title.  IV. Series

QB801.7.N52  1999        j523.8        C98-930085-4

# Contents

**1. The stars: Diamonds in the sky** ............ 4
Star tale

What is a star? ............................................ 6
Is the Sun a star?
**Try it:** Cook with the Sun's rays

How many stars are there? ........................ 8
Do other stars have planets?
Star facts

How is a star different from a planet? ..... 10
What is a shooting star?
How far away are the stars?
How do scientists measure space?

**2. Seeing stars** ............................................ 12
Orion, the great hunter

What is the Big Dipper? ........................... 14
What is the North Star?
What is the Southern Cross?
**Try it:** Make a model of a constellation

Star facts .................................................... 16
**Try it:** Make a starry slide show

Why can't we see stars during the day? ... 18
Why do stars twinkle?
Why do some stars look brighter than others?

Are the stars moving? ............................... 20
**Try it:** Check out the night sky

How do scientists study the stars? ........... 22
What is the Hubble Space Telescope?
**Try it:** Split a star's light

**3. Lives of the stars** .................................... 24
Star people

How do stars begin? .................................. 26
What is a nebula?
**Try it:** Explore Orion

How do stars grow old? ............................ 28
How do stars die?
When will the Sun die?

What is a supernova? ................................ 30
What is a black hole?
Star facts

**4. Cities of stars: Galaxies and beyond** ..... 32
Tales of the Milky Way

What is a galaxy? ...................................... 34
What is the Milky Way?
**Try it:** Reveal the secrets of the Milky Way

How many galaxies are there? .................. 36
How big is the universe?
How did the universe begin?

Will the universe ever end? ...................... 38

**Glossary** .................................................... 39

**Index** ........................................................ 40

# 1

# The stars: Diamonds in the sky

Stare up at the sky on a dark, clear night. The stars you see have dazzled sky watchers for thousands of years. Although we've learned a lot about the stars, many mysteries remain.

## Star tale

For centuries, people have told stories to explain the stars. This tale from India tells how the brightest star of all appeared in the sky.

Long, long ago, five princes set out to find the gateway to heaven. The eldest prince, Yudistira, brought his dog on the difficult journey.

One by one, the four younger princes gave up the quest and returned home. Only Yudistira hiked on, with his faithful dog trotting at his heels. He finally reached the foot of Mount Meru, where he was startled by a booming voice coming from the mountaintop.

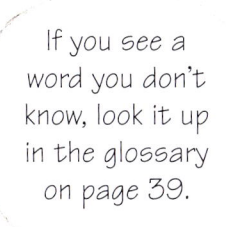

"You have reached the gateway to heaven," said the god Indra. "Enter and find total happiness."

"May I bring my dog?" asked Yudistira. "He's my one true friend."

"No dogs allowed in heaven," said Indra. "That is the rule."

"Then we're both going home," said Yudistira. He called his dog and turned back down the mountain path. They were halfway back to the village when Yudistira heard the booming voice again.

"Come back!" called Indra. "Your love for each other has changed my mind. You and the dog are welcome here."

To celebrate their arrival in heaven, Indra arranged several stars in the shape of a dog. At its heart you can still see the Dog Star, today known as Sirius. It is the brightest star that can be seen in the sky.

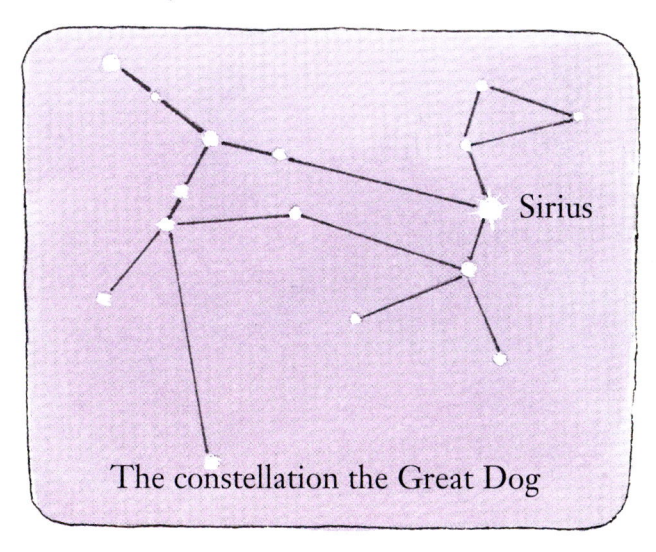

*If you see a word you don't know, look it up in the glossary on page 39.*

Sirius

The constellation the Great Dog

## What is a star?

A star is a superhot ball of gases. The center of a star is like a huge nuclear furnace. There, hydrogen gas turns into helium, giving off enormous amounts of energy. We see some of this energy as light and feel some of it as heat.

## Is the Sun a star?

Yes, our Sun is an ordinary, medium-sized star. It seems so big and bright because it is much closer to us than any other star. Earth and eight other planets circle around the Sun.

*The Sun is so big that a million Earths could fit inside it. Still, it is just a medium-sized star.*

# TRY IT!
## Cook with the Sun's rays

*You'll need:*
- a shoe box
- scissors
- aluminum foil
- tape
- a piece of coat hanger (Ask an adult to cut one for you.)
- a cooked and cooled hot dog or small sausage

1. Cut one side out of the shoe box. Ask an adult to help you poke a hole in both ends as shown.

2. Cut two ends about the size shown out of the leftover piece of box or the lid. Poke holes as shown.

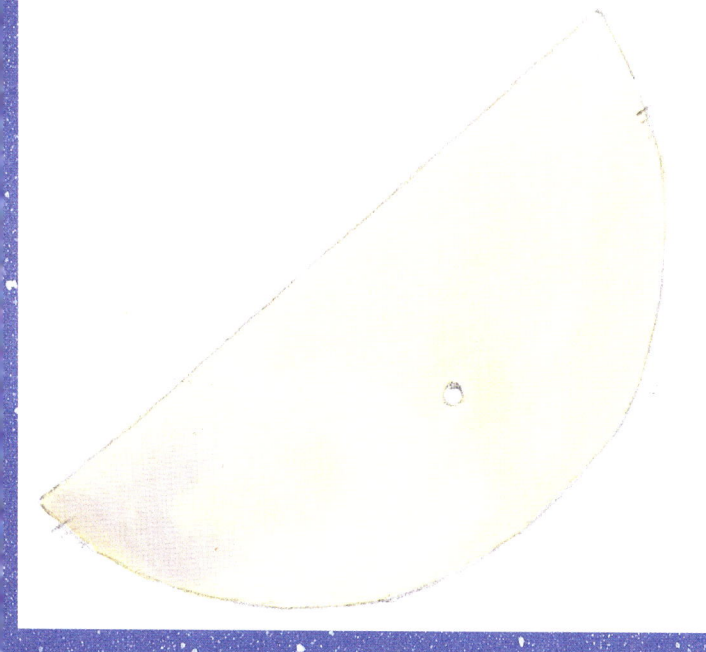

3. Attach the foil to the ends with tape. The shiny side of the foil should face in.

4. Set up your solar cooker as shown. Put it outside facing directly toward the Sun. Turn the hot dog or sausage every 10 minutes or so.

Like other stars, the Sun gives off huge amounts of energy. Some energy can be seen as light; some can be felt as heat. Does your solar cooker gather enough of the Sun's heat energy to heat up the hot dog or sausage?

**How many stars are there?**
Scientists believe that space is filled with trillions of stars. That's more than 1 000 000 000 000 stars. On a clear, moonless night far from city lights, you can see about 2000 stars.

*A scientist who studies the stars is called an astronomer. In Greek, astron means star.*

**Do other stars have planets?**
No one has actually seen planets around other stars, but there are clues that they exist. For example, astronomers have noticed that some stars seem to wobble. They think the wobbling may be caused by the pull of orbiting planets.

# STAR FACTS

All stars look white but they're not. Cooler stars are slightly red or orange, while hotter stars are blue-white. Our Sun is a medium-hot yellow star.

Stars vary in size. Some would hold millions of Suns. Others are smaller than our Moon.

Many stars come in pairs. These "binary stars" are held together by a pulling force called gravity.

The largest star you can see with your bare eyes is Betelgeuse. This supergiant star has a diameter 700 times that of the Sun.

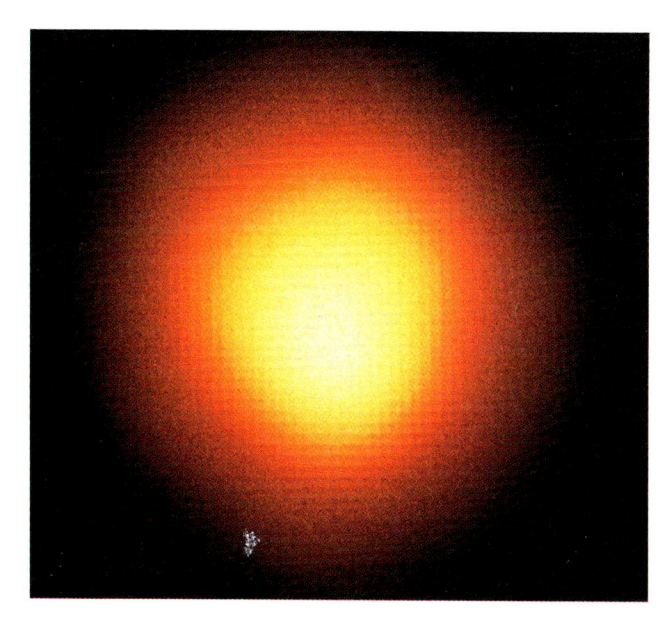

**This image of Betelgeuse was taken by the Hubble Space Telescope.**

## How is a star different from a planet?

A star glows with heat and light from the nuclear reactions in its core. Planets only appear to shine because they reflect the light from a star, our Sun.

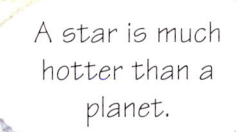

*A star is much hotter than a planet.*

## What is a shooting star?

Shooting stars aren't stars. They are meteors — brilliant streaks of light produced when particles of space dust burn up in Earth's atmosphere. A meteor particle is usually smaller than a pea. Still, it blazes a bright, fiery trail across the night sky.

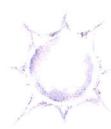

**How far away are the stars?**
Our nearest star, the Sun, is 150 million km (93 million mi.) away. That's almost next door compared to other stars. Our next closest star is Proxima Centauri. It's about 40 trillion km (24 trillion mi.) away – 280 000 times as far as the Sun.

**How do scientists measure space?**
To measure the vast distances between stars, scientists use a special unit called a light-year. This is the distance light travels in one year. Because light is so speedy, one light-year is a long, long way – about 9.5 trillion km (6 trillion mi.). Proxima Centauri, the star closest to our solar system, is 4.2 light-years away.

If a star is 100 light-years away, the light you see tonight left the star 100 years ago. So when you gaze at distant stars, you're really looking back in time.

# 2

# Seeing stars

To ancient people, the night sky was like a giant join-the-dots puzzle. They saw objects, animals, gods and people in the arrangement of the stars. Today, we call these patterns constellations. With a little practice, you too can see pictures in the sky.

This is part of the constellation Orion.

## Orion, the great hunter

The ancient Greeks named one constellation after their hero, Orion. This is the story of how Orion came to be in the sky.

Artemis, the goddess of hunting, could hit any target with an arrow. When she met the great hunter Orion, she fell deeply in love.

But Apollo, the twin brother of Artemis, was jealous. How could he get rid of his sister's new friend?

One day, Apollo saw his chance. Orion was swimming far out at sea. Apollo called his sister down to the beach.

"I bet you can't hit that black rock," he said, pointing to Orion.

"Of course I can," replied Artemis. She loaded her bow, pulled back the string and took aim. Zing! The arrow flew through the air and pierced its target. For a moment, Artemis was proud and happy. Then Apollo told her that she had just killed Orion.

Artemis broke her bow in fury. With tears running down her cheeks, she scooped Orion's body from the waves.

"You will never be forgotten," the goddess murmured as she lifted Orion into the heavens. To this day, you can spot Orion's sword, shield and belt shining in the sky.

**What is the Big Dipper?**
The Big Dipper is a pattern of seven bright stars seen in the northern hemisphere. The stars form the shape of a long-handled pot like the ones used to dip water from a pail. In Britain, people call this same star pattern the Plough.

Though it is often called a constellation, the Big Dipper is actually just one part of a constellation called Ursa Major, or the Great Bear.

**What is the North Star?**
The North Star isn't especially bright, but it is important. It seems to stand still above the North Pole. Sailors long ago used it to help them navigate.

If you live in the northern hemisphere, you can find the North Star just by looking at the Big Dipper. Just follow a line through the two stars of the Dipper's outer edge.

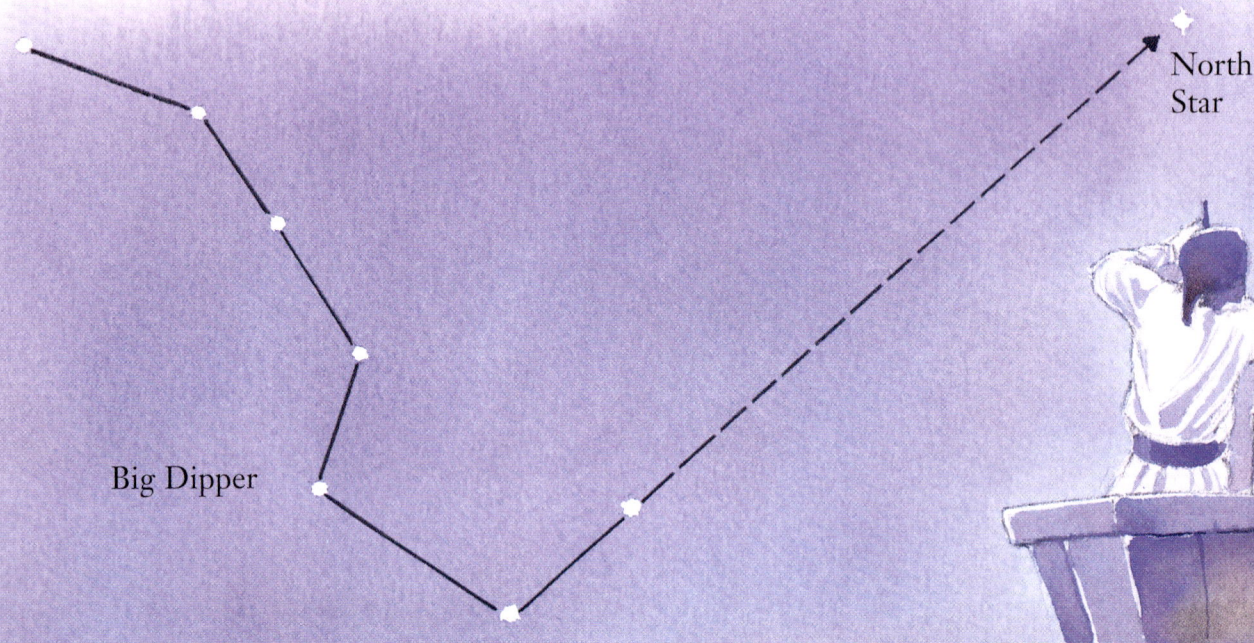

North Star

Big Dipper

**What is the Southern Cross?**
This bright, cross-shaped constellation is well-known in the southern hemisphere. Long ago sailors used the Southern Cross to help them navigate. Several countries, including Australia and New Zealand, display the Southern Cross on their flags.

Southern Cross

# TRY IT!
## Make a model of a constellation

Stars in the constellation Cygnus, the swan, form a cross-like pattern when seen from Earth. What would they look like from somewhere else in space?

*You'll need:*
- 5 wooden skewers painted black
- a ruler
- modeling clay
- aluminum foil
- a pencil and paper
- black cardboard

1. Break the skewers into the following lengths:
A: 29 cm (11 3/8 in.)
B: 19 cm (7 1/2 in.)
C: 2 cm (3/4 in.)
D: 10 cm (4 in.)
E: 30 cm (11 3/4 in.)
F: 10 cm (4 in.)

2. Stand each skewer piece in a modeling clay base. Top each with a small ball of aluminum foil. Scratch the letter for each skewer on the modeling clay base.

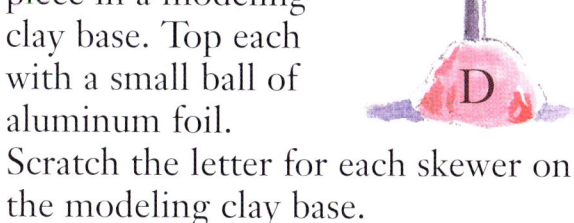

3. Lay a piece of paper over this page and trace the purple stars. Add their letters.

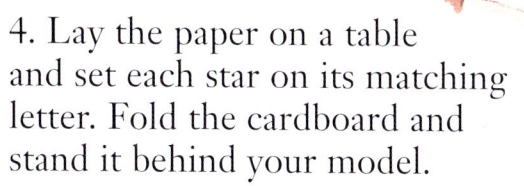

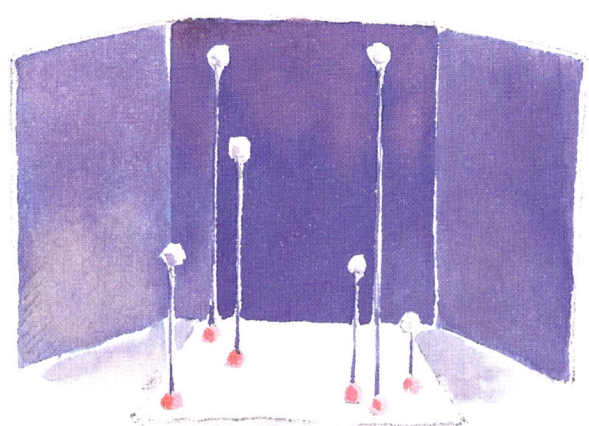

4. Lay the paper on a table and set each star on its matching letter. Fold the cardboard and stand it behind your model.

5. Look at your model from the front. Can you see a cross? Now look at your model from the side. As you can see, some stars are closer than others. In fact, some stars in Cygnus are more than 1000 light-years apart.

# STAR FACTS

Astronomers recognize 88 constellations.

Castor and Pollux, the two brightest stars in the constellation Gemini, were named after a famous pair of twins. According to Greek myth, the twin boys were the sons of Leda, queen of Sparta, and the god Zeus.

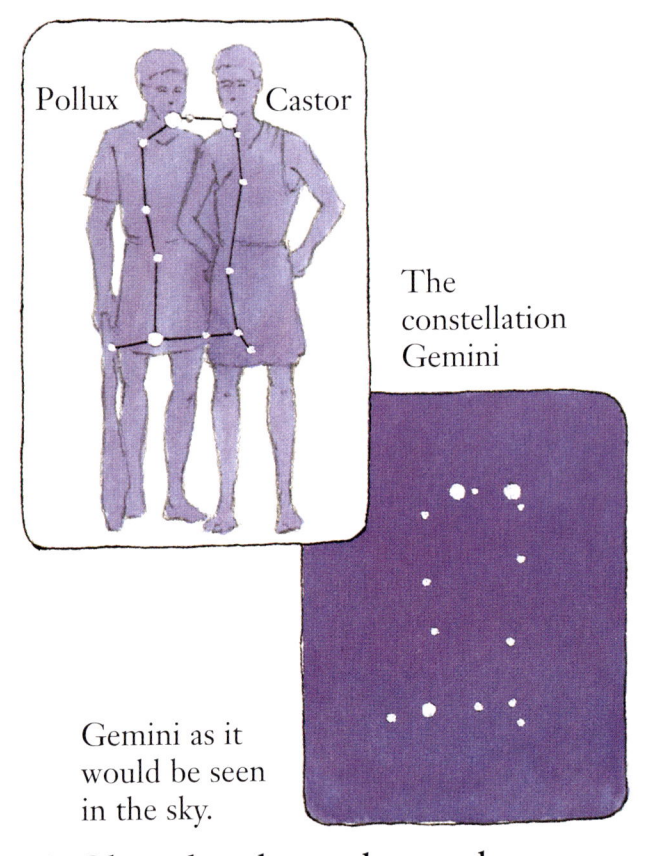

Pollux  Castor

The constellation Gemini

Gemini as it would be seen in the sky.

A Cherokee legend says the stars in the handle of the Big Dipper are a band of hunters chasing the Great Bear across the sky.

People call this constellation Leo the Lion. Does it look like a lion lying on the African plains?

For 5000 years, the constellation Taurus has been seen as a charging bull. Can you imagine the bull's horns?

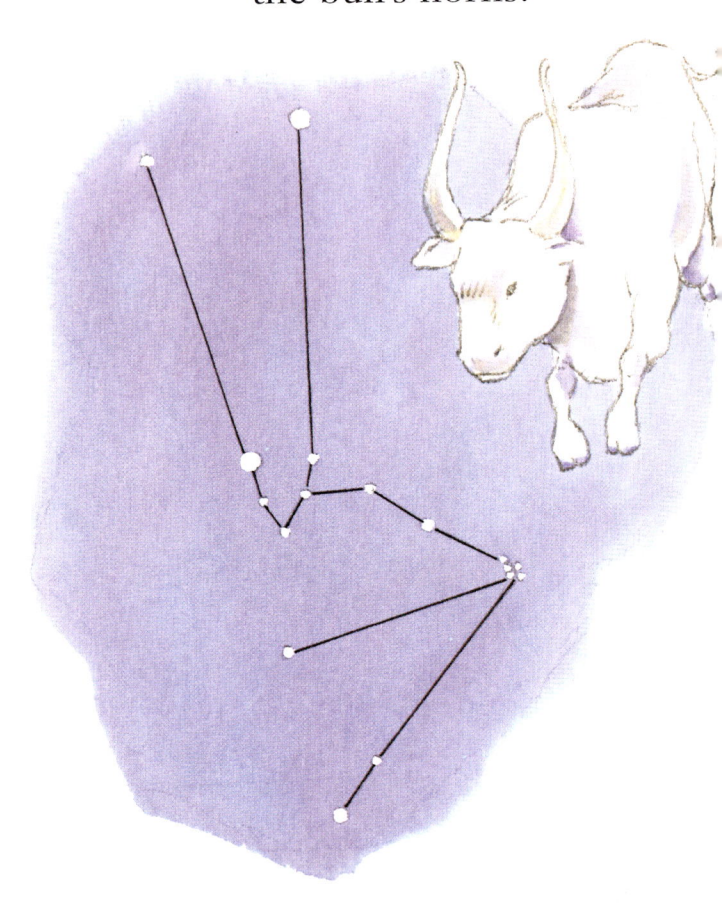

16

# TRY IT!
## Make a starry slide show

**You'll need:**
- a shoe box with a lid
- black poster paint
- a paintbrush
- scissors
- a sharp pencil
- a sheet of black bristol board
- 5 sheets of white paper
- a pin

To make the viewer:
1. Paint the inside of the shoe box and lid black.

2. Cut a rectangle out of one end of the shoe box. Leave a frame 2 cm (3/4 in.) around the sides and bottom.

3. With a pencil, pierce a peephole in the other end of the shoe box.

To make the slides:
1. Cut five rectangles of black bristol board to fit the open end of the shoe box. Cut five pieces of white paper the same size.

2. Draw five constellations on the white paper. Use the constellations from the inside cover of this book as a guide.

3. Glue one constellation drawing onto each black rectangle.

4. With the pin, poke a hole for each star in the constellation. For larger stars, make bigger holes.

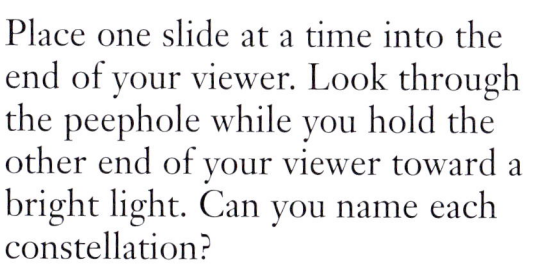

Place one slide at a time into the end of your viewer. Look through the peephole while you hold the other end of your viewer toward a bright light. Can you name each constellation?

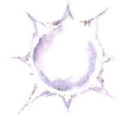

**Why can't we see stars during the day?**
Stars are always there in the sky. We can't see them during the day because the Sun's light makes the sky too bright. The same thing happens if you shine a flashlight in a bright room – you can't see the beam of light. But in the dark, there it is, shining brightly.

**Why do stars twinkle?**
The light from a star travels a long way to reach Earth. By the time it gets close to Earth, it is like a narrow beam. Particles in Earth's atmosphere, the blanket of air around our planet, bounce the light around. This makes the star appear to twinkle.

**Why do some stars look brighter than others?**
Some stars look bright because they're big. Others look bright because they're closer to us. Scientists call a star's brightness its magnitude. *Apparent* magnitude tells how bright the star looks from Earth. *Absolute* magnitude tells how bright the star really is compared to other stars.

**Are the stars moving?**
The stars seem to move slowly across the sky every night. Watching them, you might think that the stars are circling around us. But they're not – they only appear to move because the Earth is spinning.

Also, because the Earth travels around the Sun each year, the stars seem to change position from season to season. Actually, Earth's travels just give us a different view of the sky in summer, fall, winter and spring. So we can't see any real movement in the stars from day to day or month to month.

However, if we could watch the stars over many thousands of years, changes in their positions would be noticeable. That's because the stars are actually traveling through space.

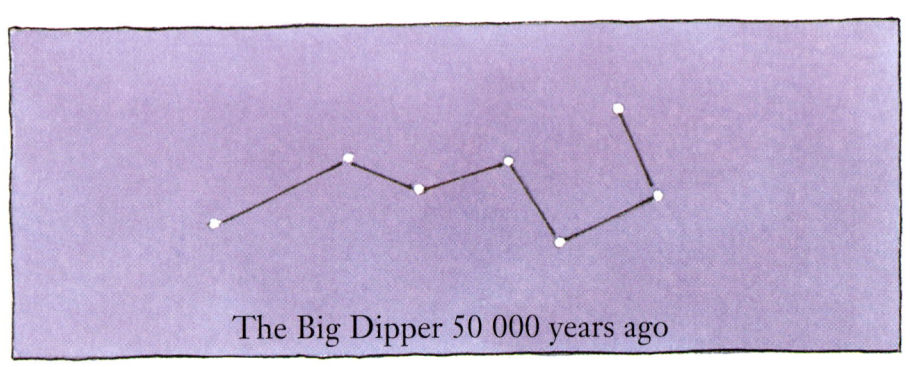

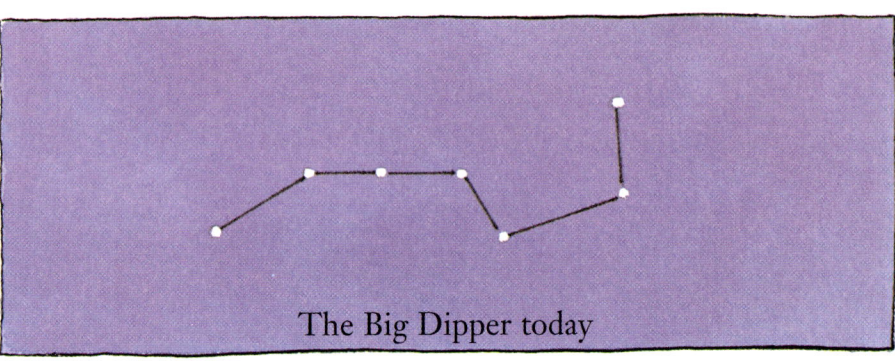

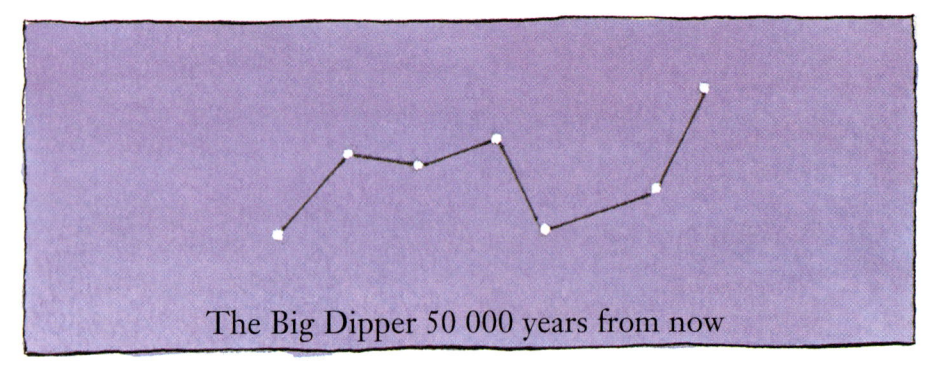

The Big Dipper 50 000 years ago

The Big Dipper today

The Big Dipper 50 000 years from now

20

# TRY IT!
## Check out the night sky

**You'll need:**
- a clear night
- an open space away from bright lights
- an old blanket
- a flashlight with the light covered with red tissue paper
- the constellation pictures from the inside cover of this book

1. Spread your blanket on the ground.

2. Use the flashlight to check the constellation pictures. The red tissue paper helps keep your eyes adjusted to the dark.

3. Look for the brightest stars in the sky. These stars are usually found in major constellations. Remember that the major constellations cover large areas of the sky.

Constellations are in different parts of the sky at different times of the year. To find out what you can expect to see this month, check the star charts in an astronomy magazine or on the Internet.

## How do scientists study the stars?

Astronomers use telescopes on the Earth and in space to study the stars. They study the stars' visible light (the light we can see) and invisible light energy (such as radio waves and infrared radiation we can't see).

The light from each star can be split into a pattern called a spectrum. By studying a star's spectrum, scientists can figure out how hot the star is, what it is made of and how fast it is moving.

## What is the Hubble Space Telescope?

Imagine looking up at the sky from the bottom of a lake. That's pretty much what happens when we observe the stars through Earth's atmosphere. To get a clearer picture, scientists have sent telescopes into orbit beyond Earth's atmosphere.

The most famous of these is the Hubble Space Telescope. It was launched into orbit in 1990. At first, Hubble didn't work very well because of problems with its mirrors. Then, in 1993, it was repaired during a special space mission. Since then, the Hubble has been sending incredible star portraits back to Earth.

The Infrared Astronomical Satellite (IRAS) was put into orbit in 1983. It picks up infrared energy, a part of starlight we can't see.

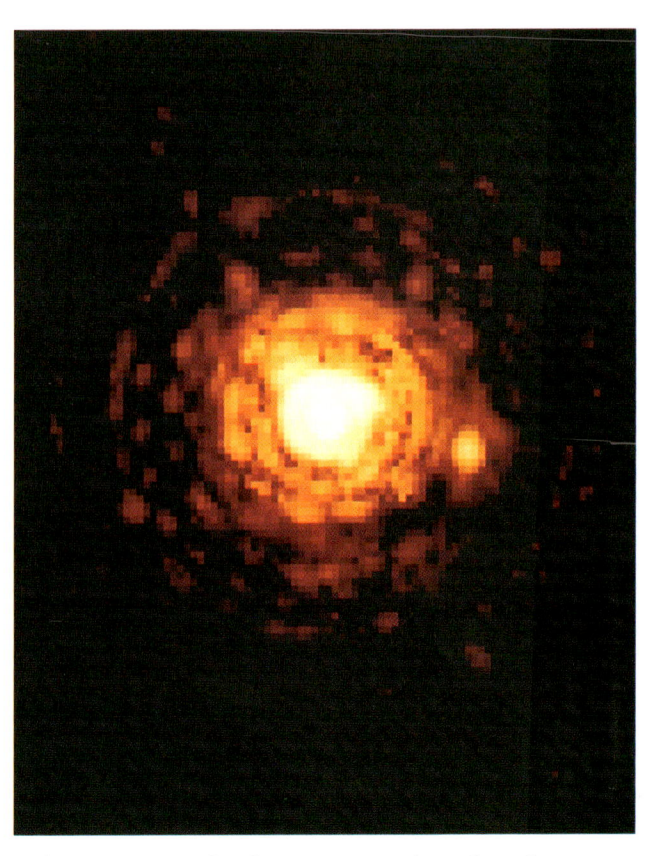

This image of Gliese was taken by the Hubble Space Telescope. Gliese is one of the smallest stars in the Milky Way galaxy — 60,000 times fainter than our Sun.

# TRY IT!
## Split a star's light

**You'll need:**
- a small mirror
- a clear wide-mouthed drinking glass

1. Set the mirror inside the glass so that it is resting at an angle.

2. Fill the glass with water.

3. Place the glass and mirror in a sunny window.

4. Move the glass until the mirror reflects sunlight onto the ceiling. You will see a rainbow of colors.

**Sunlight looks white but it is really a mixture of colors.**

The colors you see are the visible part of the spectrum of our nearest star, the Sun. By studying its spectrum, astronomers can investigate a star.

# Lives of the stars

Stars don't last forever.
Like us, they are born, grow old and die.
In early times, people didn't know about the stages of a star's life. Even so, they told stories about a Sky World where Star People lived and died.

A star is formed in the constellation Orion.

## Star people

The Micmac people of eastern Canada tell this legend.

One warm, clear night, two sisters lay on their fur sleeping robes under the stars.

"I wish I could marry that small, red star," said Younger Sister.

"I'd choose that gorgeous white one," replied Older Sister. "He's so bright and shiny." Soon both sisters were fast asleep.

Younger Sister awoke with a jolt. A little old man with red eyes was standing beside her. Frightened, she poked her sister.

"What's the matter?" asked Older Sister groggily. Then she gasped. Beside her stood a handsome young man with shining eyes.

"We are going hunting now," said the young man. "Take care of the camp and gather firewood. But do not lift that flat rock by the big tree."

As soon as the two men were gone, Younger Sister ran to the flat rock. She lifted it, stared into the hole below and screamed. Older Sister came running.

Both sisters stared into the hole. Far, far below they could see forests, lakes and rivers.

"We are in the Sky World," Older Sister said in amazement.

"Then those two men must be our Star Husbands," said Younger Sister. "Our wishes have come true!"

25

## How do stars begin?

Stars form out of clouds of gas and dust particles in space. As the particles swirl around, some of them clump together. More and more particles join the clump. Then, under the force of its own gravity, the clump begins to shrink inward. Intense pressure makes the core of the star hotter and hotter. Finally, it becomes hot enough to set off constant nuclear reactions. A star has been born.

## What is a nebula?

Any cloud of gas and dust in space is called a nebula. Some nebulae glow with the light of stars within them. Others are dark and block our view of the stars behind them. Many nebulae are called "star nurseries" because new stars form from their gases.

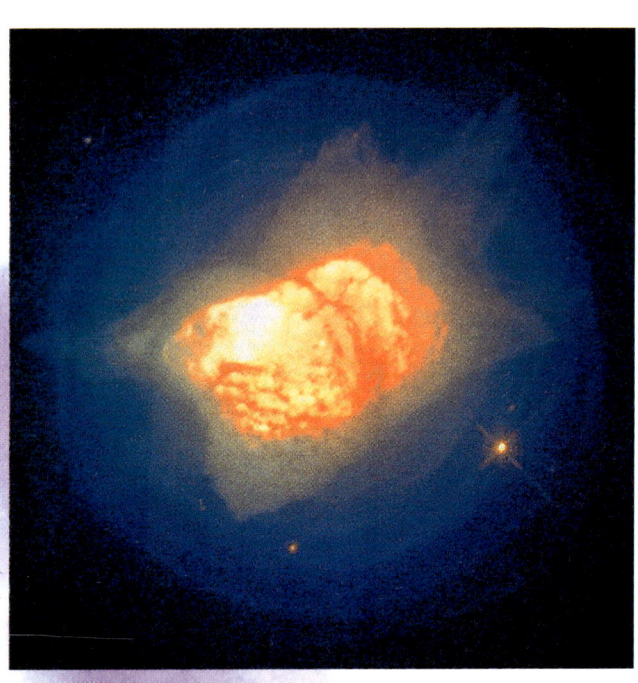

**This picture of a nebula shows clouds of red dust surrounding a hot center.**

# TRY IT!
## Explore Orion

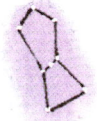

**You'll need:**
- a clear, moonless night in late fall or winter
- a pair of binoculars

1. Find the constellation Orion in the sky. Look for his belt of three bright stars in a straight line.

2. Find Betelgeuse, a red supergiant star, in Orion's shoulder. Betelgeuse is 1400 light-years away from Earth.

3. Look for Rigel, a blue supergiant star, on Orion's knee. Rigel is also about 1400 light-years from us.

4. With your binoculars, take a close look at the hazy-looking "star" near the tip of Orion's sword. This bright patch is actually the Orion nebula, a gigantic cloud in which new stars are being born. Though the nebula looks small, it is about 16 light-years across.

In the southern hemisphere, Orion appears to be hanging upside down.

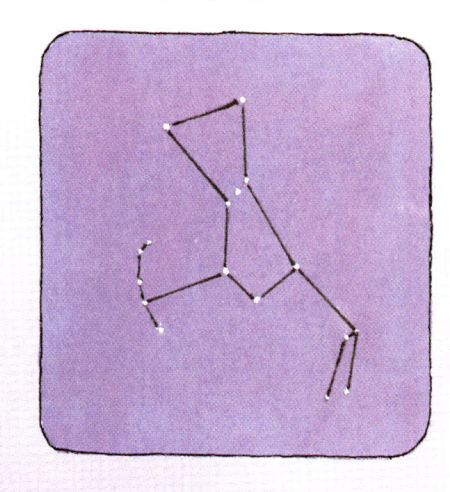

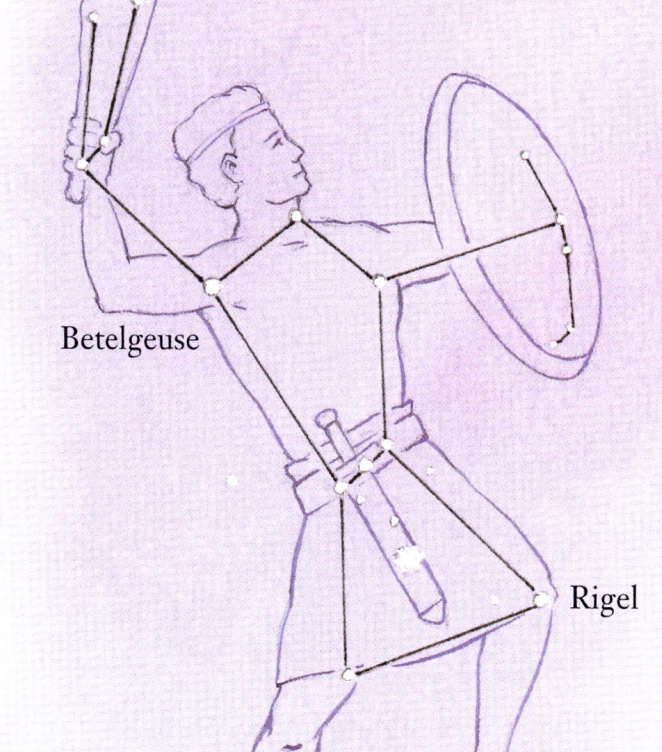

**How do stars grow old?**
A star spends most of its life glowing steadily. During this time, it uses hydrogen for fuel. Then, after billions of years, the hydrogen in the core runs low. That's when the star enters old age. It begins to burn the hydrogen in the shell around the core. This change makes the star expand and change color. From a mid-sized yellow or white star, it grows into a red giant.

A mid-sized yellow star

A red giant

## How do stars die?

After it becomes an elderly red giant, a star continues to change. It may expand and contract as it builds up and loses outer layers of glowing gases. Finally, the old star blasts off all its red-hot gases. It collapses into a small star called a white dwarf. White dwarf stars are usually about the size of the Earth. Because they are so dense, they're very heavy. A white dwarf the size of Earth would weigh as much as the Sun.

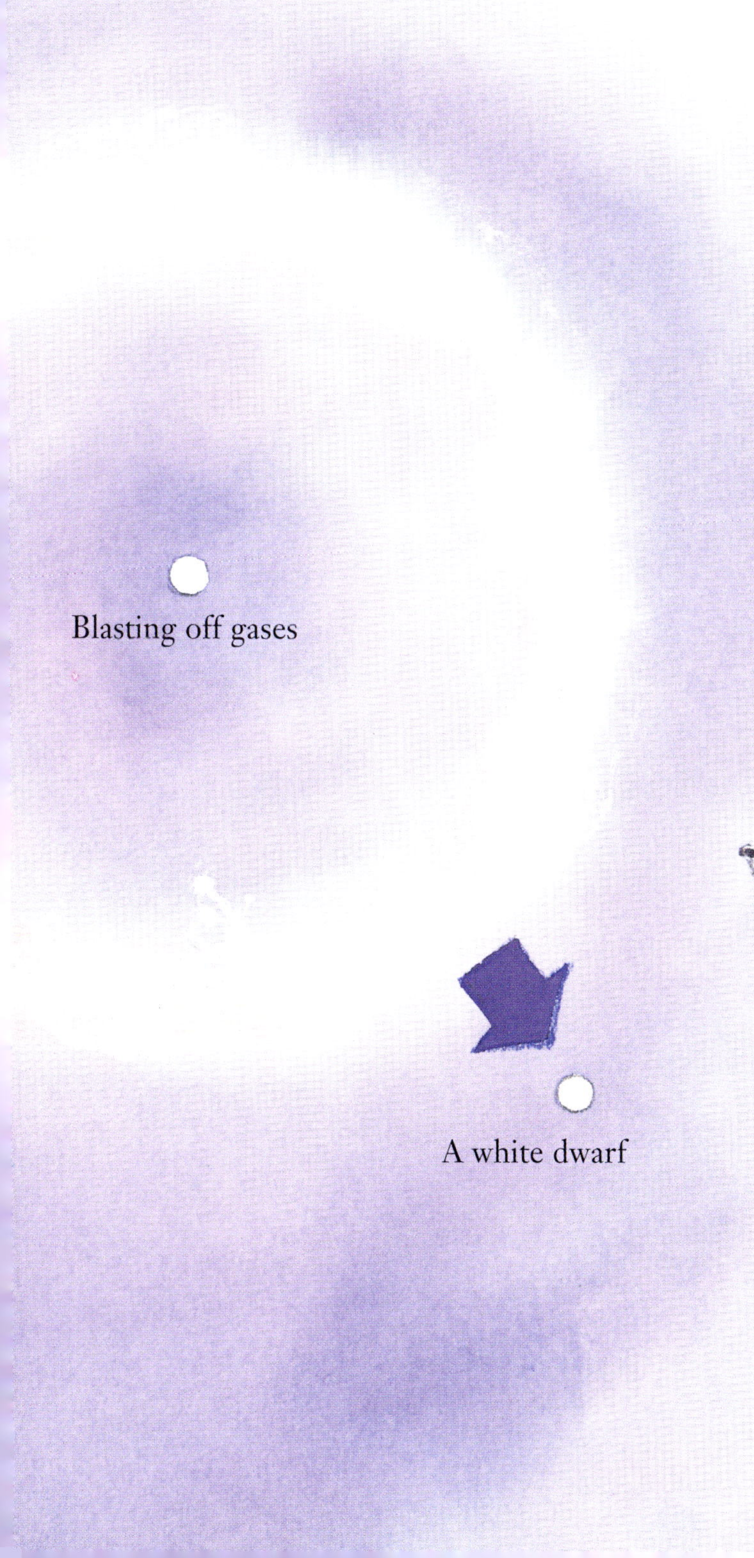

Blasting off gases

A white dwarf

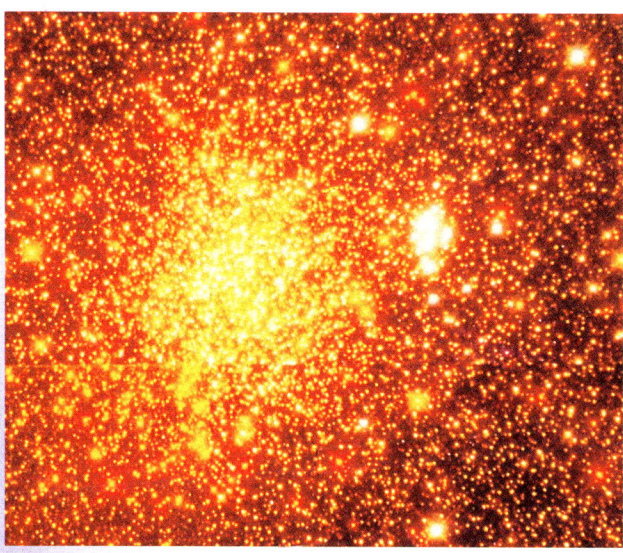

**This star cluster contains a mixture of young white and yellow stars and older red giants.**

*A spoonful of a white dwarf star would weigh as much as four full-grown elephants.*

## When will the Sun die?

The Sun is a middle-aged star. It will probably shine brightly for about 5 billion more years. After that, it will swell into a red giant and finally die as a white dwarf.

**What is a supernova?**
A supernova is the final explosion of a dying supergiant star. After the main part of its life is over, a really huge star (like Rigel or Betelgeuse) expands into a supergiant. Then, under the force of its own powerful gravity, it begins to collapse. It collapses so quickly that it explodes in a tremendous burst of light. One supernova can be brighter than a whole galaxy.

> When ancient stars exploded, bits of star material shot out into the universe. Some of it swirled into a cloud of particles that eventually formed our Sun and its planets, including Earth. Everything on Earth – trees, tigers, mountains and people – has been formed from that original stardust. You could say, you're a star!

## What is a black hole?
Astronomers think that black holes form when the very largest, brightest stars, called blue supergiants, collapse at the end of their lives. A black hole is a huge amount of matter and energy squished into a very small place. Because a black hole is so dense, its gravity is incredibly strong. Nothing – not even light – can escape from its pull. So no one has ever seen one. But still, they're believed to exist.

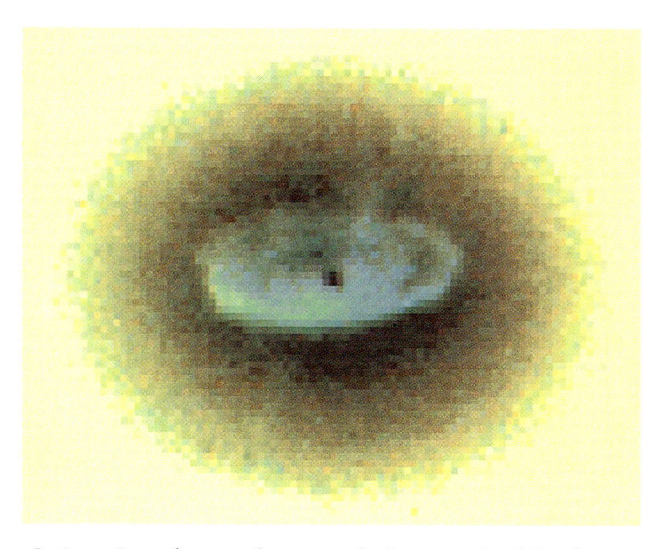

**Scientists have detected signs of a black hole at the center of this far-off galaxy. The cloud of dust surrounding it is being sucked into it.**

## STAR FACTS

Mira, a star in the constellation Cetus, appears to grow bright, fade and grow bright again every few months. This happens because Mira is expanding and contracting in its old age.

Canadian astronomer Ian Shelton spotted a bright supernova in 1987. The last time anyone had seen one like it was in 1604.

The Pleiades, or Seven Sisters, is a cluster of bright young stars in the constellation Taurus. These stars were "born" about 70 million years ago, when the last dinosaurs roamed the Earth.

# Cities of stars: Galaxies and beyond

On a clear, dark night, far from city lights, look for a pale band of light stretching across the sky. It is the Milky Way galaxy, home to millions of stars including our Sun. But ancient people invented stories to explain this ribbon of light.

This infrared image shows the central part of the Milky Way galaxy.

## Tales of the Milky Way

The Vikings saw the Milky Way as the road to Valhalla, the home of the gods.

Native North Americans said that the white trail was snow shaken from the back of Grizzly Bear as it climbed its way into the heavens.

In eastern Europe, Estonians told of Lindu, a young goddess who was engaged to the Northern Lights. Lindu prepared for the wedding, but her groom never arrived. Lindu still weeps in the heavens with her white bridal veil drifting across the sky.

In China, people told of a young weaver who married a cowherd. One day the weaver was called back into the heavens by her uncle, the Yellow Emperor. Her desperate young husband tried to follow her but was stopped by a wide white river. The cowherd and the weaver are two bright stars in the summer sky. They still gaze at each other across the river of brightness we call the Milky Way.

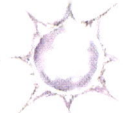

## What is a galaxy?
A galaxy is a vast collection of stars. Galaxies also contain gas and dust, all held together by gravity. Galaxies come in different shapes and sizes. Large galaxies hold billions of stars. Some look like misty blobs, but others have definite shapes that look like hot dogs or pinwheels.

The Hubble Space Telescope took this picture, which showed several hundred galaxies that had never before been seen.

## What is the Milky Way?
The Milky Way is the galaxy we live in. It spins through space like an enormous pinwheel made of stars.

Our Sun is one of about 200 billion stars in this vast galaxy. Light from a star on the far side of the Milky Way would take about 100 000 years to reach Earth.

The name Milky Way is also given to the starry band of light we see in the sky when we look toward the center of the galaxy.

We Earthlings are moving through space in one of the arms of the Milky Way.

# TRY IT!
## Reveal the secrets of the Milky Way

*You'll need:*
- a clear, moonless night
- an open viewing place away from city lights
- binoculars

1. Gaze at the whole sky. Every star you see is part of the Milky Way galaxy. Earth is located in one of the spiral arms of the disk-like galaxy.

2. Find a hazy band of light across the sky. Now you're looking deep into a neighboring arm of the Milky Way, toward the center of the galaxy.

3. Look at the same hazy band with your binoculars. Surprise! It's made up of millions of stars.

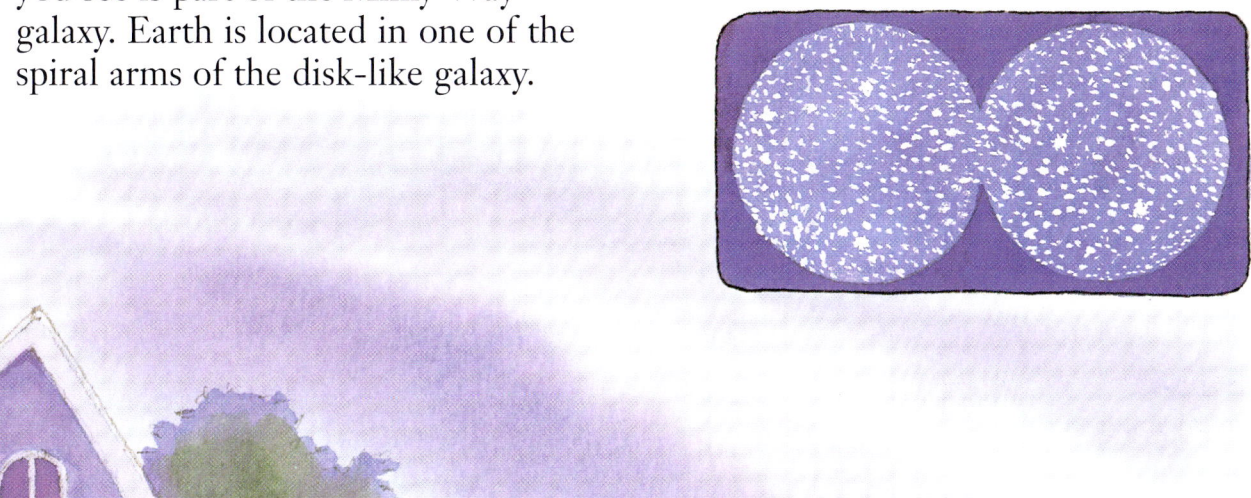

**How many galaxies are there?**
The universe has billions of galaxies. Astronomers once believed that the galaxies were scattered evenly through space, but we now know that they are clumped together in clusters and huge superclusters.

The Milky Way galaxy is part of a cluster of about 30 galaxies that astronomers call the Local Group.

This "cartwheel galaxy" formed when two galaxies collided. The blue ring is made up of millions of newly formed stars.

**How big is the universe?**
The universe is bigger than anyone can imagine. Astronomers have picked up radio signals from a galaxy 12 billion light-years away. That's incredibly far, but scientists believe the universe goes even farther. Some galaxies might be so far away that their light still hasn't reached us. In fact, many astronomers think that the universe has no edges.

**How did the universe begin?**
No one knows how the universe first came to be, but scientists have figured out some things about its past. Early on, the universe was incredibly hot, and everything in it was squeezed close together in a space smaller than a grain of sand. Then, about 10 or 15 billion years ago, the universe suddenly expanded and cooled. Dust and other particles spread rapidly through space. Scientists call this process the Big Bang.

Over millions of years, the spreading particles clumped together, forming galaxies, stars and planets. Even today, they continue to spread apart. The universe is still expanding.

**Will the universe ever end?**
Some astronomers believe that the universe is like an elastic band that can stretch only so far before it starts to pull in again. If this is the case, the universe may shrink back into a tiny point, heat up, and then start again with another Big Bang. The whole process would take billions of years.

Most astronomers, on the other hand, think the universe will continue to expand forever. Either way, you should wonder, not worry, about the universe. It's going to be around for a long, long time.

# Glossary

**astronomer:** someone who studies the stars, planets and other objects in space

**atmosphere:** a layer of gases surrounding a planet

**Big Bang:** the sudden expansion of the early universe from a tiny, hot lump of matter

**binary stars:** pairs of stars held together by gravity

**black hole:** an extremely dense object with gravity so strong it traps everything nearby, including light

**constellation:** a group of stars that can be seen as a pattern in the sky

**galaxy:** a vast collection of stars, gas and dust held together by gravity

**gravity:** an invisible pulling force that pulls objects in the universe toward one another. For example, sometimes two stars can be held together by this force.

**infrared radiation:** a form of light energy we can't see

**light-year:** the distance light travels in one year

**magnitude:** the brightness of a star or other object in space

**meteor:** a flash of light from a burning particle falling through Earth's atmosphere

**Milky Way:** the galaxy in which we live

**nebula:** a cloud of gas and dust in space

**planet:** a large object that orbits a star and does not make its own light. Earth is a planet that orbits a star called the Sun.

**radio waves:** an invisible form of energy

**red giant:** an old star that has grown to many times its original size

**shooting star:** another name for a meteor

**spectrum:** the pattern formed when starlight is split

**star:** a ball of burning gases that gives off light

**supergiant:** a very large, very bright star

**supernova:** the brilliant final explosion of a dying supergiant star

**telescope:** an instrument that makes faraway objects seem nearer. Telescopes are often used to look at the stars.

**universe:** everything that exists, including billions of galaxies

**white dwarf:** the small, dense star left behind when a red giant loses its outer layers

# Index

activities
    constellation model, 15
    cooking with Sun's rays, 7
    explore Orion, 27
    show Sun's spectrum, 23
    sky watching, 21
    star slide show, 17
    view Milky Way, 35
aging of stars, 28-30

Betelgeuse, 9, 27
Big Bang, 37, 38
Big Dipper, 14, 20
binary stars, 9
birth of a star, 26
black hole, 31

Cetus, 31
color of stars, 9
constellations, 12-17
    Big Dipper, 14, 20
    Cetus, 31
    Cygnus, 15
    Gemini, 16
    Leo, 16
    Orion, 12, 13, 24, 27
    Plough, 14
    Southern Cross, 14
    Taurus, 16
    Ursa Major, 14
Cygnus, 15

death of a star, 29
distance to stars, 11, 36
Dog Star, 5

galaxies, 32-37
Gemini, 16
Gliese, 22

helium, 6
Hubble Space Telescope, 9, 22, 34
hydrogen, 6, 28

Infrared Astronomical Satellite (IRAS), 22

Leo, 16
light-years, 11
Local Group, 36

magnitude, 19
meteors, 10
Milky Way, 32, 33, 34, 35
Mira, 31
motion of stars, 20

nebula, 26, 27
North Star, 14
number of galaxies, 36
number of stars, 8

Orion, 12, 13, 24, 27

planets, 8, 10
Pleiades, 31
Plough, 14
Proxima Centauri, 11

red giant star, 28, 29
Rigel, 27

Seven Sisters, 31
Shelton, Ian, 31
shooting stars, 10
Sirius, 5
sky watching, 21
solar cooker, 7
Southern Cross, 14
spectrum, 22, 23
stories about stars, 5, 13, 16, 25, 33
Sun, 6, 7, 9, 11, 29
supergiant, 30
supernova, 30

Taurus, 16
telescopes, 22
twinkling, 18

universe, 36, 37
Ursa Major, 14

white dwarf, 29

40